YOUR KNOWLEDGE HAS VALUE

- We will publish your bachelor's and
  master's thesis, essays and papers

- Your own eBook and book -
  sold worldwide in all relevant shops

- Earn money with each sale

Upload your text at www.GRIN.com
and publish for free

Bibliographic information published by the German National Library:

The German National Library lists this publication in the National Bibliography;
detailed bibliographic data are available on the Internet at http://dnb.dnb.de .

Imprint:

Copyright © 2019 GRIN Verlag
Print and binding: Books on Demand GmbH, Norderstedt Germany
ISBN: 9783668985803

This book at GRIN:

https://www.grin.com/document/492920

Bojidarka Ivanova, Michael Spiteller

On the temperature dependence on the stochastic dynamic mass spectrometric diffusion parameter

Stochastic dynamic mass spectrometry

GRIN Verlag

ON THE TEMPERATURE DEPENDENCE ON THE STOCHASTIC DYNAMIC MASS SPECTROMETRIC DIFFUSION PARAMETER

PREFACE

The major aim of this work is to introduce our new model equation connecting among the so–called *stochastic dynamic diffusion coefficient* "D$_{SD}$," the experimental mass spectrometric outcome *intensity* "I" and the experimental parameter *temperature* "T," respectively. A closer review of our contributions, so far, to the domain of the *"stochastic dynamic mass spectrometry"* has shown that we have developed a functional relationship between the D$_{SD}$ parameter and the mass spectrometric intensity. It is $$D_{SD} = 1.3193 \times 10^{-14} \times A \times \frac{\left(\overline{I^2} - \left(\overline{I} \right)^2 \right)}{\left(I - \overline{I} \right)^2}$$. Its universal applicability to a set of soft–ionization mass spectrometric methods has been evidenced within a small–scale research on correlation between theory and experiment, which has been tested by chemometrics. As a corollary, there has been concluded that the temporal behaviour of the experimental mass spectrometric intensity obeys a certain law and this law is the equation shown above. The fundamental differences in the functional relationship written before and our new innovative model lies in that we account for the effect of the temperature on the D$_{SD}$ parameter and the experimental mass spectrometric measurable variable. The roots of the latter model are again to stochastic plausibility theories, focusing the attention on the Gillespie's exact numerical solution of the Ornstein–Uhlenbeck process according to the forward Fokker–Planck equation (or the forward Kolmogorov equation) and the theory of continuous Markov processes. We shall not only introduce, a new functional relation, but also we shall provide persuasive empirical proofs of that the new formula is true. The contribution explores our own experimental mass spectrometric data.

The discussion, herein, provides sufficient justification, that the content of the work would be of interest in MSc students specializing in *"Advanced methods for the Analytical Chemistry"* or *"Environmental Chemistry;"* students specializing in *"Theoretical and computational chemistry;"* and PhD students or researchers developing the mass spectrometric methodology.

TABLE OF CONTENT

ACKNOWLEDGMENTS

The authors thank the Deutsche Forschungsgemeinschaft, supporting their study in this research topic within a project grant 255/22–1; the Alexander von Humboldt Stiftung; the Deutscher Akademischer Austausch Dienst for a grant within priority program "Stability Pact South-Eastern Europe" and for purchasing on Evolution 300 UV–VIS–NIS spectrometer; and central instrumental laboratories for structural analysis at Dortmund University of Technology as well as analytical and computational laboratory clusters at the Institute of Environmental Research at the same University.

The work was carefully carried out. Nevertheless, authors and publisher do not warrant the information therein to be free of errors. The work is being published in English aiming at a widest access to the scientific contributions. English is not native language of the authors, however. Stylistic rough edges might occur. The authors hope for understanding of the reader.

Conflicts of interest

Michael Spiteller has received research grant (255/22–1, DFG); Bojidarka Ivanova has received research grant (255/22–1, DFG).

Address correspondence to the authors:

Lehrstuhl für Analytische Chemie, Institut für Umweltforschung, Fakultät für Chemie und Chemische Biologie, Universität Dortmund, Otto–Hahn–Straße 6, 44221 Dortmund, Nordrhein–Westfalen, Deutschland.

KEYWORDS

Mass spectrometry; stochastic dynamic diffusion; temperature

CHAPTER 1

ABSTRACT

The work is devoted to a new stochastic dynamic equation, providing correlation among the so–called *stochastic dynamic diffusion parameter* "D_{SD}," the temperature "T" and the experimental mass spectrometric variable *intensity "I,"* respectively. The functional relation is:

$$\ln\left(\overline{(I-\bar{I})^2}\right) = -\ln\left(-\left(\ln\left(\frac{k_B \times T}{m}\right)\right)^3 \times \frac{2 \times \Delta t \times T \times k_B}{m \times D_{SD}}\right)$$

. The experimental proof of its validity has been obtained on the base on electrospray ionization, matrix–assisted laser desorption ionization, atmospheric pressure chemical ionization and collision induced dissociation mass spectrometry of eight analytes and their fifty seven fragment ions. The experimental design encompasses a temperature range T = 273.15–723.15 K and spans of scan times Δt = 1.86–19.98 s.

INTRODUCTION

The major aim of this commentary is to discuss our innovative strategy based on stochastic dynamics (SD) applicable to quantify the experimental mass spectrometric (MS) outcome *intensity* [1–8], in particular, looking at our new model equation (2). Because of, in the light of the accumulated experimental facts, the already developed by us model equation (1) connecting the D_{SD} parameter with the MS intensity appears a governing law determining the temporal behaviour of the MS intensity with respect to the short spans of the scan time. In other words, the functional relation between the D_{SD} parameter of a MS ion and the intensity of its observable MS peak obeys a certain law and this law is namely equation (1). In order to, simplify the presentation of our theory we will concentrate only on basic concepts behind the later model relationship. There have been adopted the stochastic dynamic Box–Müller method and Ornstein–Uhlenbeck formalism as well as the Einstein's approximation to the Ornstein–Uhlenbeck process [1–8]. As can be seen from the results presented in the later references our criterion of evidencehood is the empirical testability of equation (1). There have been examined electrospray ionization (ESI), atmospheric pressure chemical ionization (APCI), collision induced dissociation (CID) and matrix–assisted laser desorption ionization (MALDI) mass spectra of organic and metal–organic analytes of $Ag^{I}-$, $Zn^{II}-$ and $Cu^{II}-$ions, respectively. In this context, a logic question is: what is the motive to derive the new equation (2)? A central problem for addressing to this new model is the quantitative relation among not only between D_{SD} parameter and the MS intensity values, but also among the latter parameters and one of the most important experimental factors, among others, determining the ionization efficiency of analyte ions, *i.e.* the *temperature* [9–22]. In order to, gain an evidential respectability for our study the equation (2) is tested on analytes, which D_{SD} parameters have been published [1–9]. Thus, the validity of equation (1) has been documented. In this context, we only open to debate, herein, the validity of the new equation (2). To draw persuasive conclusion about its justification we examine both theoretically and experimentally eight analytes and their fifty seven fragment MS ions over spans of scan time $\Delta t = 1.86$–19.98 s and temperature range T = 273.15–723.15 K. Herein, we mainly discuss the validity and universal applicability of equation (2) to low temperatures, in spite of, results from heated ESI– (HESI) and APCI–MS measurements are presented, as well. Its application to higher temperatures has been tested on a set of statistically representative organics, so far, using HESI and APCI methods at T = 573.15–723.15 K [23]. The reliability of the model applied to the later temperature range shows coefficients of linear correlation between theory and experiment $r = 0.9904$–0.99939 (HESI) and $r = 0.9959$–0.99977 (APCI) studying five configurationally locked polyenes and their [2+2] cycloaddition products of interaction and, in total, fifty seven analyte MS ions [23]. The latter evidences and arguments are obviously sound. In the previous work [23] we have examined the validity of equation (2) using two independent MS methods at higher temperatures, while the purpose of the current study is to demonstrate its validity again looking at MS results from two independent methods (ESI and MALDI), however, at lower experimental temperatures. The evidential link between the two works is that they provide a complete picture of the universal applicability not only to equation (1), but also to equation (2) looking at a statistically representative set of analytes and their fragment MS ions within a large range of temperatures (T = 273.15–723.15 K) and four soft–ionization MS methods, for instance, ESI, APCI, MALDI and CID, respectively.

$$D_{SD} = 1.3193 \times 10^{-14} \times A \times \frac{\left(\overline{I^2} - \left(\overline{I} \right)^2 \right)}{\left(I - \overline{I} \right)^2}$$

(1).

2. THEORY

The derivation of model equation (2) adopts the Gillespie's exact numerical solution of the Ornstein–Uhlenbeck process [24]. It is based on the forward Fokker–Planck equation or the forward Kolmogorov equation [25–29] within the theories of a continuous Markov process [25–28] and the Einstein's formulas [30,31] of the characteristic functions A(x,t) and D(x,t) (equations (3) and (4)) of the probability density function shown by means of the forward Fokker–Planck equation in the Gillespie's original work [24]. According to the Gillespie's theory, the variance (σ^2) is given by equation (5), where k_B is the Boltzmann constant ($1.3806.10^{-23}$ $m^2.kg.s^{-2}.K^{-1}$); $\Delta t = t - t_0$, T – temperature [K]; m – mass and D – diffusion coefficient [$cm^2.s^{-1}$], but according to the Einstein's formula [29,30], respectively.

$$\ln\left(\overline{(I-\bar{I})^2}\right) = -\ln\left(-\left(\ln\left(\frac{k_B \times T}{m}\right)\right)^3 \times \frac{2 \times \Delta t \times T \times k_B}{m \times D_{SD}}\right) \tag{2}$$

$$A(x,t) = \frac{k_B \times T}{D \times m} \tag{3}$$

$$D(x,t) = \frac{2}{D} \times \left(\frac{k_B \times T}{m}\right)^2 \tag{4}$$

$$\sigma^2 = \frac{k_B \times T}{m} \times \left(1 - e^{-\frac{2 \times \Delta t \times k_B \times T}{D \times m}}\right) \tag{5}$$

By contrast, Gillespie's theory [24], in writing equation (2) we only partially adopt the Einstein's formulas to A(x,t) and D(x,t) functions. Equation (3) is kept with a remark that "D" denotes the SD diffusion parameter according to our equation (1). The second function D(x,t) we modify empirically and write its as equation (6).

$$D(x,t) = \frac{-2 \times \ln\left(\frac{k_B \times T}{m}\right)^3 \times k_B \times T}{D_{SD} \times m} \tag{6}$$

In equation (6) the D_{SD}, again, denotes *diffusion coefficient*, but according to equation (1).

At this point, let us add few comments in defence of our innovative view of the last equation (6). First, the characteristic functions A(x,t) and D(x,t) of an Ornstein–Uhlenbeck process are designed to account for drift and diffusion functions [24,28]. The form of the former function, however, has been taken as spring submerged in a fluid with high viscosity [32]. However, our remark on the inapplicability of equation (4) in [23] does address namely the nature of the ESI–MS continuum. It cannot be regarded as a liquid having high viscosity. Second, within the fluctuation-dissipation relation the two functions reflecting drift and diffusion are mutually connected [24,28,32]. Therefore, in choosing form of the discussed arguments [33] there is important to take into consideration the fact that their monotonic mutual functional relationship should be affected. According to Gillespie's exact numerical solution of the Ornstein–Uhlenbeck process within the aforementioned approximations to drift and fluctuation the mutual relationship between the functions A(x,t) and D(x,t) is given by equation (7) [24,28].

$$D(x,t) = -\frac{2 \times k \times T}{x} \times A(x,t)$$

(7).

According to our choice of a form of D(x,t), the mutual relationship between the discussed characteristic functions becomes equation (8), whereas mentioned before the drag coefficient is kept, while the coefficient accounting for the fluctuating force is empirically modified.

$$D(x,t) = -2 \times \ln\left(\frac{k \times T}{m}\right)^{3} \times A(x,t)$$

(8).

3. RESULTS

Figures 1–6 are drawn from MS results from the following analytes: metal–organic Cu^{II}–complex of the amino acid glycine, $\{[Cu^{II}(gly)_2.2H_2O]\}$ [6]; the amino acid L–tryptophyl–L–tryptophan (H–Trp–Trp–OH) [3]; the organic dye (4-diethylamino-benzylidene)-pyridin-4-yl-amine [1] and organic salt 1-(2-chloro-ethyl)-pyrrolidinium 2-hydroxy-5-sulfo-benzoate [5]; glycyl containing homopenta and homohexaoligopeptides (G5 and G6) [4]; a Zn^{II}–complex of G5 [8], nitroamine derivative of 3-aminomethyl-3,5,5-trimethyl-cyclohexylamine (NA) [4], and cytidine [2] respectively. As far as, our usual route of structural analysis includes MS and crystallographic analyses of the analytes, the latter references have detailed on experimental MS ions together with fragment paths of the analytes, in addition to, information about single crystal X–ray diffraction. The results in this work will shift the focus of the MS analysis from determination of D_{SD} parameters according to equation (1) to quantitative relations among the parameter and values of I and T according to the new model equation (2). Importantly, the new fact that we are endeavouring to underline in this section is that the coefficients of statistical correlation obtained according to this new functional relation are excellent: $r = 0.9865_8$–0.99947_7. This fact, directly evidences not only the validity of the model, but also its broad applicability to different ionization methods, temperatures and type of analytes, respectively. The following $\ln\left(\overline{\left(I-\bar{I}\right)^2}\right)$ values are used to **Figure 1**: 19.0280_8 (*m/z* 117), 16.8751_9 (*m/z* 119), 19.7896_4 (*m/z* 183), 16.3158_6 (*m/z* 185), 26.0261_8 (*m/z* 156), 24.3190_8 (*m/z* 158), 23.3452_9 (*m/z* 171) and 21.5314_3 (*m/z* 173) $(\{[Cu^{II}(gly)_2.H_2O]\})$, while the

$$-\ln\left(-\ln\left(\frac{k_B \times T}{m}\right)^3 \times \frac{2 \times T \times \Delta t \times k_B}{m \times D_{SD}}\right)$$

values for the shown ions are: 19.0895_9, 17.1032_1, 20.6027_7, 17.0852_9, 26.4660_2, 24.8519_8, 23.8151_3 and 22.0866_3, respectively; the $\ln\left(\overline{\left(I-\bar{I}\right)^2}\right)$ values are: 20.0284 (*m/z* 285), 19.3175_5 (*m/z* 287), 11.4461_4 (*m/z* 134), 10.8399_8 (*m/z* 136), 25.1884 (*m/z*

303) and 23.3894 (m/z 305) (1-(2-chloro-ethyl)-pyrrolidinium 2-hydroxy-5-sulfo-benzoate),

while the $-\ln\left(-\ln\left(\frac{k_B \times T}{m}\right)^3 \times \frac{2 \times T \times \Delta t \times k_B}{m \times D_{SD}}\right)$ values for the shown ions are: 20.4774, 17.5585,

11.5851₂, 11.2828₉, 26.6356₇ and 24.0653₆, respectively. The corresponding data used to

Figure 3 are $\ln\left(\overline{(I-\overline{I})^2}\right)$ values: 31.5821 (m/z 361), 26.9309 (m/z 247), 25.9548 (m/z 201),

26.5702 (m/z 190), 24.1688 (m/z 172), 22.6039 (m/z 133), 22.7576 (m/z 115) and 17.419 (m/z

109) (G6) while the $-\ln\left(-\ln\left(\frac{k_B \times T}{m}\right)^3 \times \frac{2 \times T \times \Delta t \times k_B}{m \times D_{SD}}\right)$ values for the shown ions are: 32.3421,

27.8478, 27.6113, 28.1423, 24.7289, 23.9356, 22.784 and 17.5423, respectively. The

$\ln\left(\overline{(I-\overline{I})^2}\right)$ values used to the same figure are: 23.4143 (m/z 115), 22.0413 (m/z 247), 21.453

(m/z 259) and 30.9489₉ (m/z 252) (G5), while the $-\ln\left(-\ln\left(\frac{k_B \times T}{m}\right)^3 \times \frac{2 \times T \times \Delta t \times k_B}{m \times D_{SD}}\right)$ values for

the shown ions are: 22.8121, 21.7731, 21.3752 and 31.1502, respectively. The chemometric

analysis shown in **Figure 4** encompasses MS results from Zn^{II}–complexes of G5 according to

the data in reference [8]. There are studied two molar ratios metal–to–ligand (M:L): M:L =

1:1 and 1:4. The $\ln\left(\overline{(I-\overline{I})^2}\right)$ values used to the latter figure are: 21.2099 (m/z 387), 21.3518

(m/z 441), 21.8049 (m/z 450), 26.9618 (m/z 304) and 26.9772 (m/z 326) (M:L = 1:1), while

the $-\ln\left(-\ln\left(\frac{k_B \times T}{m}\right)^3 \times \frac{2 \times T \times \Delta t \times k_B}{m \times D_{SD}}\right)$ values for the shown ions are: 22.2766, 22.54018, 22.9998,

26.96512 and 27.1636; and the $\ln\left(\overline{(I-\overline{I})^2}\right)$ values used to the figure are: 21.6155 (m/z 387),

21.8747 (m/z 441), 22.0775 (m/z 450), 29.4021 (m/z 304) and 28.5506 (m/z 326) (M:L = 1:4),

while the $-\ln\left(-\ln\left(\frac{k_B \times T}{m}\right)^3 \times \frac{2 \times T \times \Delta t \times k_B}{m \times D_{SD}}\right)$ values for the shown ions are: 23.0172, 22.9626,

23.1908, 30.0014 and 29.3018, respectively. The corresponding data taken from reference [4] about the analyte nitroamine derivative of 3-aminomethyl-3,5,5-trimethyl-cyclohexylamine (NA), which are quantify according to equation (2) and depicted in **Figure 5** are as followings: the $\ln\left(\overline{\left(I-\bar{I}\right)^2}\right)$ values are 28.3911 (*m/z* 106), 24.6013 (*m/z* 131), 24.1504 (*m/z* 171), 23.6804 (*m/z* 154), 25.5021 (*m/z* 154), 23.4433 (*m/z* 137), 18.0319 (*m/z* 214), 23.5541 (*m/z* 184), 19.8378 (*m/z* 159), 18.9312 (*m/z* 199) and 20.5083 (*m/z* 137), while the $-\ln\left(-\ln\left(\dfrac{k_B \times T}{m}\right)^3 \times \dfrac{2 \times T \times \Delta t \times k_B}{m \times D_{SD}}\right)$ values are 28.0037, 24.0976, 24.0418, 23.549, 25.9766, 23.2617, 17.9052, 24.2708, 19.9069, 19.3842 and 20.5556, respectively.

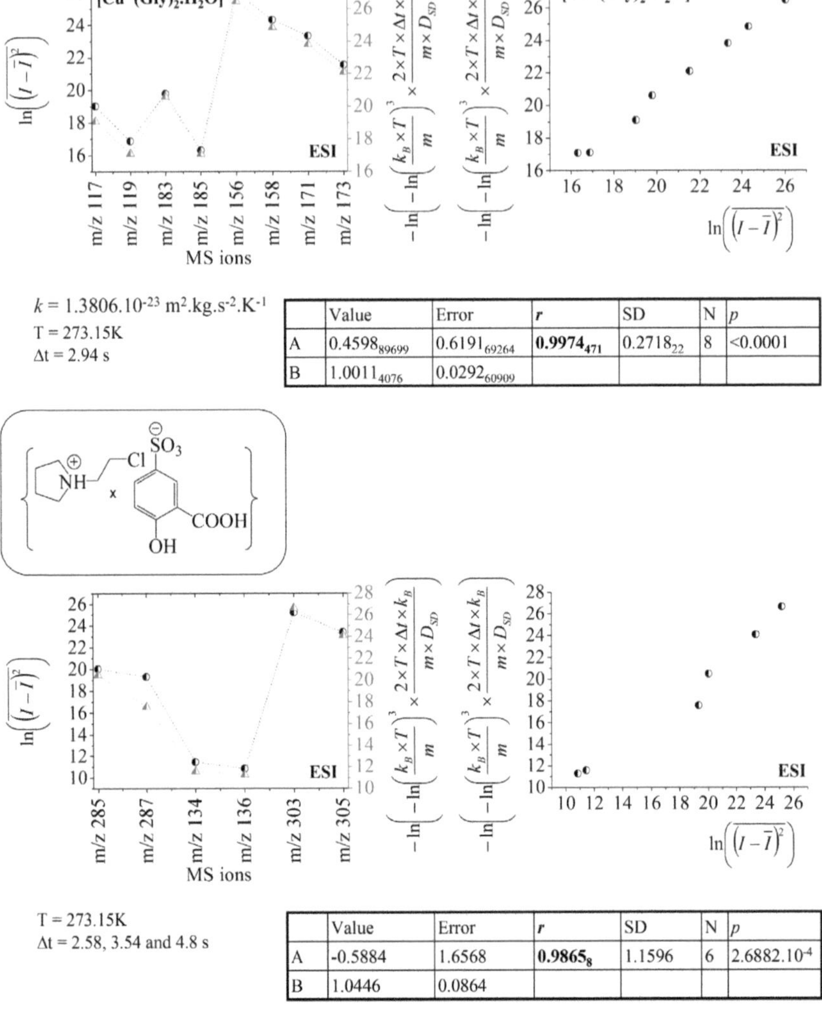

$k = 1.3806.10^{-23}\ \mathrm{m^2.kg.s^{-2}.K^{-1}}$

$T = 273.15\,\mathrm{K}$

$\Delta t = 2.94\ \mathrm{s}$

	Value	Error	r	SD	N	p
A	0.4598_{89699}	0.6191_{69264}	$\mathbf{0.9974_{471}}$	0.2718_{22}	8	<0.0001
B	1.0011_{4076}	0.0292_{60909}				

$T = 273.15\,\mathrm{K}$

$\Delta t = 2.58,\ 3.54\ \mathrm{and}\ 4.8\ \mathrm{s}$

	Value	Error	r	SD	N	p
A	-0.5884	1.6568	$\mathbf{0.9865_8}$	1.1596	6	$2.6882.10^{-4}$
B	1.0446	0.0864				

Figure 1. Relations between $\ln\left(\overline{(I-\bar{I})^2}\right)$ and $-\ln\left(-\ln\left(\dfrac{k_B \times T}{m}\right)^3 \times \dfrac{2 \times T \times \Delta t \times k_B}{m \times D_{SD}}\right)$ of MS ions of analytes $\{[Cu^{II}(gly)_2.H_2O]\}$ and 1-(2-chloro-ethyl)-pyrrolidinium 2-hydroxy-5-sulfo-benzoate; chemometrics; temperatures and scan times (Δt) [s]; chemical diagram of the latter analyte.

The $\overline{\left(I-\bar{I}\right)^2}$ and D_{SD} [cm^2.s^{-1}] values are according to [5,6]; the synthesis of the analtes was carried out according to [5,6].

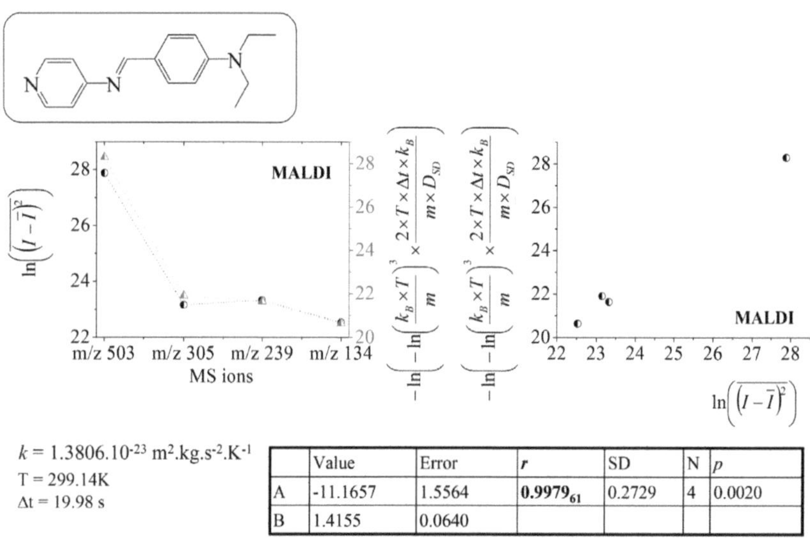

$k = 1.3806.10^{-23}$ m^2.kg.s^{-2}.K^{-1}
$T = 299.14$K
$\Delta t = 19.98$ s

	Value	Error	r	SD	N	p
A	-11.1657	1.5564	0.9979_{61}	0.2729	4	0.0020
B	1.4155	0.0640				

Figure 2. Relations between $\ln\left(\overline{\left(I-\bar{I}\right)^2}\right)$ and $-\ln\left(-\ln\left(\frac{k_B \times T}{m}\right)^3 \times \frac{2 \times T \times \Delta t \times k_B}{m \times D_{SD}}\right)$ of MS ions of analyte (4-diethylamino-benzylidene)-pyridin-4-yl-amine; chemometrics; the $\ln\left(\overline{\left(I-\bar{I}\right)^2}\right)$ values are: 27.8767_2 (*m/z* 503), 23.1463_6 (*m/z* 305), 23.3238 (*m/z* 239) and 22.532 (*m/z* 134), while the $-\ln\left(-\ln\left(\frac{k_B \times T}{m}\right)^3 \times \frac{2 \times T \times \Delta t \times k_B}{m \times D_{SD}}\right)$ values for the shown ions are: 28.2912, 21.9073, 21.6365 and 20.6383, respectively; chemical diagram; the $\overline{\left(I-\bar{I}\right)^2}$ and D_{SD} [cm^2.s^{-1}] data are according to [1]; the synthesis of the analyte was carried out according to [1].

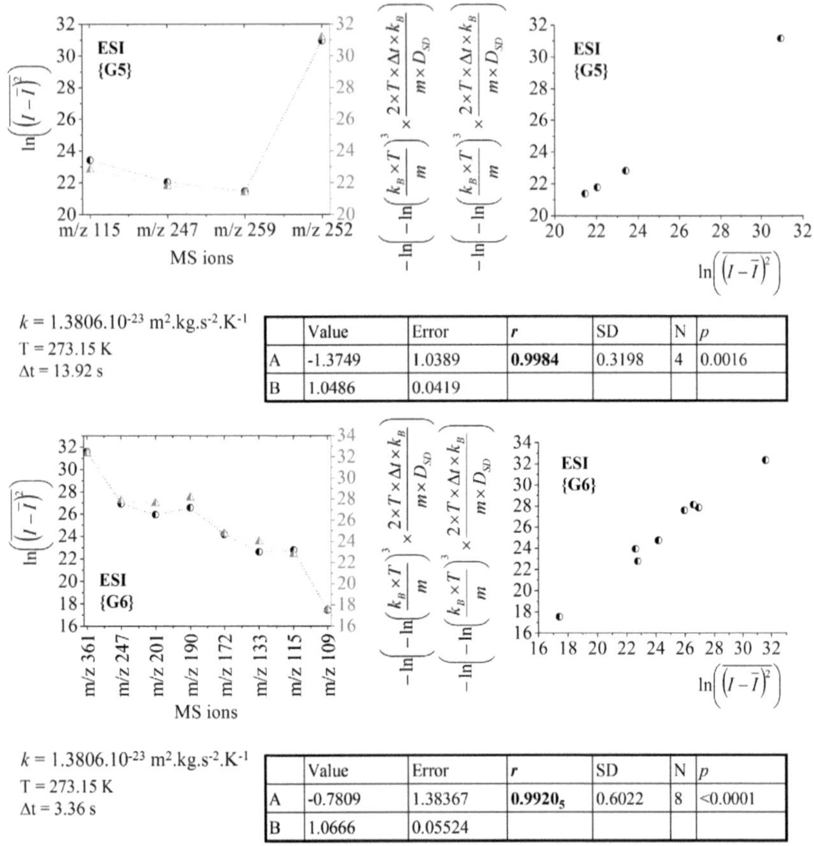

Figure 3. Relations between $\ln\left(\overline{(I-\bar{I})^2}\right)$ and $-\ln\left(-\ln\left(\frac{k_B \times T}{m}\right)^3 \times \frac{2 \times T \times \Delta t \times k_B}{m \times D_{SD}}\right)$ of MS ions of analytes

G5 and G6; chemometrics; temperature and scan times (Δt) [s]; the $\overline{(I-\bar{I})^2}$ and D_{SD} [cm^2.s^{-1}]

values are according to [4]; the synthesis of the analytes was carried out according to [4].

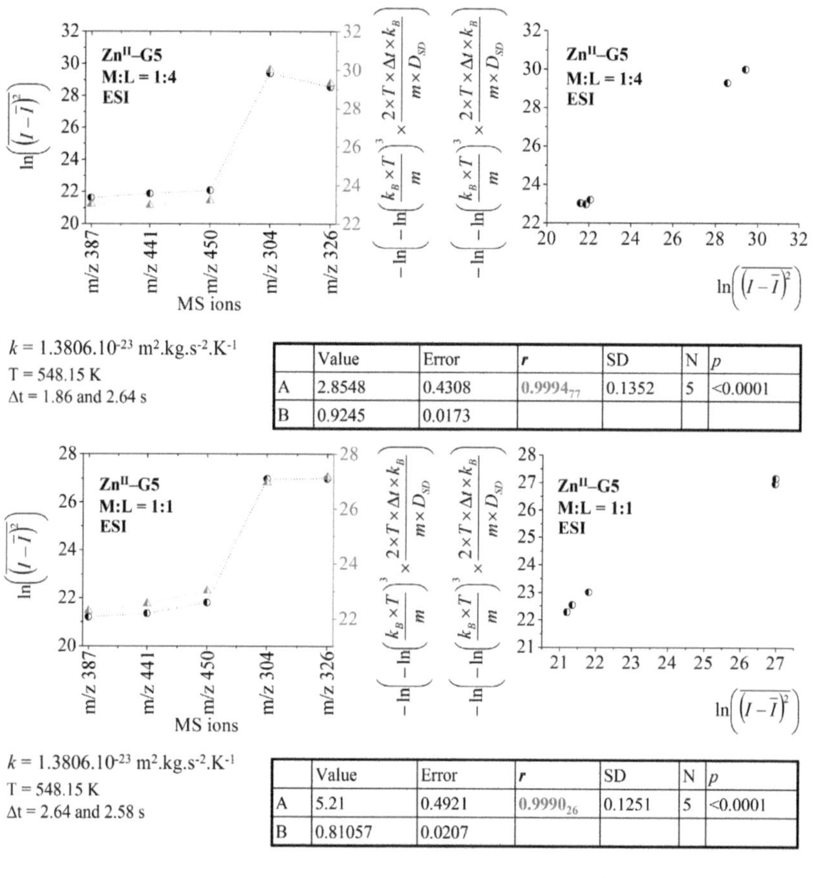

$k = 1.3806 \cdot 10^{-23}$ m^2.kg.s^{-2}.K^{-1}
T = 548.15 K
Δt = 1.86 and 2.64 s

	Value	Error	r	SD	N	p
A	2.8548	0.4308	0.9994$_{77}$	0.1352	5	<0.0001
B	0.9245	0.0173				

$k = 1.3806 \cdot 10^{-23}$ m^2.kg.s^{-2}.K^{-1}
T = 548.15 K
Δt = 2.64 and 2.58 s

	Value	Error	r	SD	N	p
A	5.21	0.4921	0.9990$_{26}$	0.1251	5	<0.0001
B	0.81057	0.0207				

Figure 4. Relations between $\ln\left(\overline{(I-\bar{I})^2}\right)$ and $-\ln\left(-\ln\left(\frac{k_B \times T}{m}\right)^3 \times \frac{2 \times T \times \Delta t \times k_B}{m \times D_{SD}}\right)$ of MS ions of analytes Zn^{II}–G5; chemometrics; temperature and scan times (Δt) [s]; the $\overline{(I-\bar{I})^2}$ and D$_{SD}$ [cm^2.s^{-1}] values are according to [8]; the synthesis of the analytes was carried out according to [8].

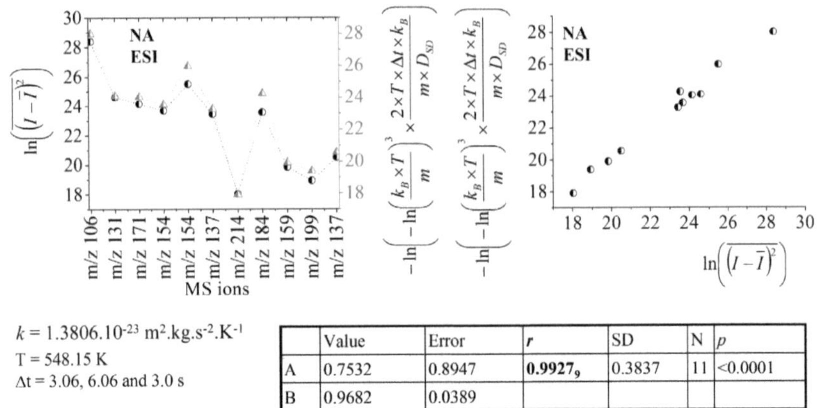

$k = 1.3806.10^{-23}$ m^2.kg.s^{-2}.K^{-1}

T = 548.15 K

Δt = 3.06, 6.06 and 3.0 s

	Value	Error	*r*	SD	N	*p*
A	0.7532	0.8947	**0.9927**$_9$	0.3837	11	<0.0001
B	0.9682	0.0389				

Figure 5. Relations between $\ln\left(\overline{(I-\bar{I})^2}\right)$ and $-\ln\left(-\ln\left(\dfrac{k_B \times T}{m}\right)^3 \times \dfrac{2 \times T \times \Delta t \times k_B}{m \times D_{SD}}\right)$ of MS ions of analyte (NA); chemometrics; temperature and scan times (Δt) [s]; the $\overline{(I-\bar{I})^2}$ and D$_{SD}$ [cm^2.s^{-1}] values are according to [4]; the synthesis of the analytes was carried out according to [4].

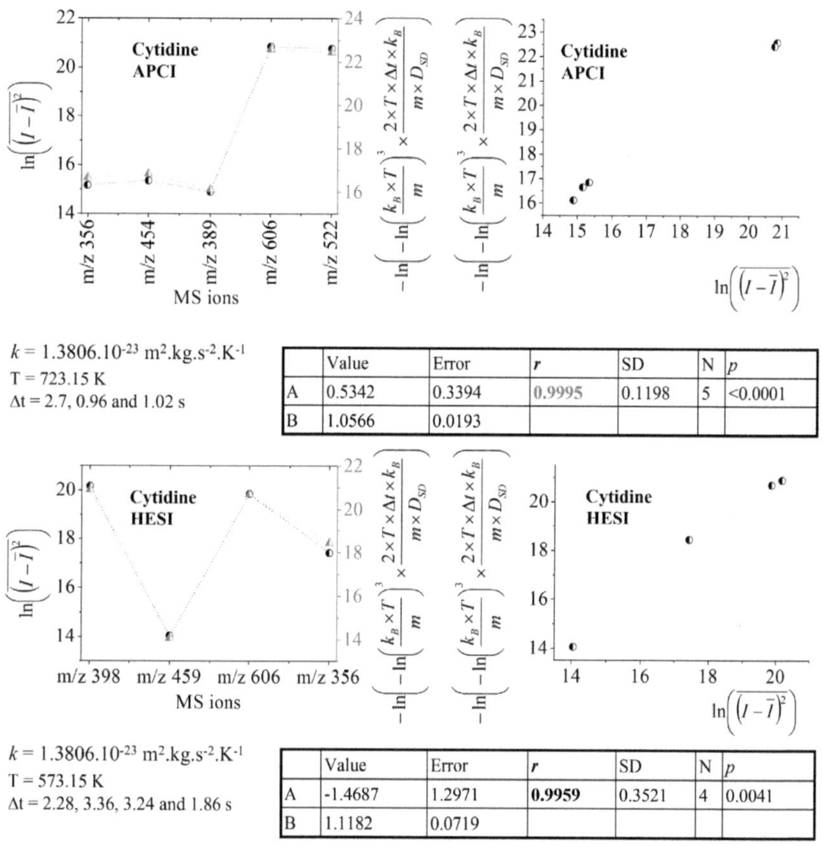

$k = 1.3806.10^{-23}$ m^2.kg.s^{-2}.K^{-1}
T = 723.15 K
Δt = 2.7, 0.96 and 1.02 s

	Value	Error	*r*	SD	N	*p*
A	0.5342	0.3394	0.9995	0.1198	5	<0.0001
B	1.0566	0.0193				

$k = 1.3806.10^{-23}$ m^2.kg.s^{-2}.K^{-1}
T = 573.15 K
Δt = 2.28, 3.36, 3.24 and 1.86 s

	Value	Error	*r*	SD	N	*p*
A	-1.4687	1.2971	**0.9959**	0.3521	4	0.0041
B	1.1182	0.0719				

Figure 6. Relations between $\ln\left(\overline{(I-\bar{I})^2}\right)$ and $-\ln\left(-\ln\left(\dfrac{k_B \times T}{m}\right)^3 \times \dfrac{2 \times T \times \Delta t \times k_B}{m \times D_{SD}}\right)$ of MS ions of analyte cytidine; chemometrics; temperature and scan times (Δt) [s]; the $\overline{(I-\bar{I})^2}$ and D$_{SD}$ [cm^2.s^{-1}] values are according to [2].

4. DISCUSSION

In this section we will engage in an amount of discussion of our theoretical SD proposal puts forward in **section 3** dealing with *"theory."* As aforementioned, in this work we are going to elaborate new equation allowing for a quantitative treatment of the mass spectrometric intensity with respect to the measurement time, but accounting for the effect of the temperature. An enormous amount of effort has gone into direction to interconnect among different SD concepts. So far, in our research contributions devoted to apply SD methods to quantify the MS outcome *"intensity"* [1] we have introduced the D_{SD} parameter connecting between property of analyte ion with experimental absolute intensity of the MS peak of the same species. As mentioned before, a statistically significant experimental evidences have been accumulated supporting for the claim that equation (1) appears universally applicable approach to different soft–ionization methods, experimental conditions and type of the analytes. With the new model equation (2) presented and discussed in this work, we do not attempt to diminish the significant importance of equation (1) to treat quantitatively the experimental MS intensity. On the contrary, there have been presented empirical facts that the D_{SD} parameter provides direct and exact multidimensional structural information about the ions – 3D molecular conformation and electronic structure – when experimental ultra–high accuracy MS analyses are employed complementary with high accuracy quantum chemical methods in obtaining the so–called *quantum chemical diffusion parameter* "D_{QC}" within the Arrhenius's approximation [1–9]. On this base, one might go onto arise the following question: What light shed the new model equation (2) onto the correlation among the experimental factor *temperature*, the MS outcome *intensity* and the molecular level aspects of the multidimensional conformation and electronic structures of the ions? A question "why the Einstein's formula of $D(x,t)$ cannot be applied directly to ESI–MS data?" has already been approached to reference [23]. Given that, there is needed to throw light upon the former question. For us, the key contribution of equation (2), explicitly, to the *structural analytical chemistry* – its great applicability to the *quantitative analytical chemistry* is self–evident – is that, namely the new model provides direct quantitative link among the D_{SD} parameter, the intensity and the temperature, respectively. Depending on the temperature, there is able to examine exactly by means of model equations (1) and (2) the multidimensional structural changes of the analytes with respect to the experimental factor *temperature via* a correlation between D_{SD} and D_{QC} parameters as has been illustrated previously [1–9]. At this point, we would like to add and important remark on the capability of the mass spectrometry as robust analytical instrumentation in a general context. It is useful to remind the readers of this paper that the current view of the capability of the mass spectrometry is as an irreplaceable analytical instrumentation, but chiefly of quantitative analysis. Even, references [34,35] have claims, as we can see, that the MS methods are inapplicable to multidimensional structural analysis. We should like to underline, that through our contributions, so far, [1–9] we have opposed the later view providing empirical proofs of an excellent statistical coefficients of correlation between D_{SD} and D_{QC} parameter, which demonstrates persuasively that the soft–ionization MS methods are able to provide, in fact, exact 3D molecular conformational and electronic structure of the analytes. The following coefficients of linear correlation between the lastly discussed parameters have been obtained: $r = 0.9806_8$ [7], 0.9833, 0.9901 and 0.9806_8–0.9956 [4] as well as 0.95575 [3], respectively. These excellent chemometric data clearly reject the statements in works [33,34] of the boundaries of applicability of the mass spectrometry to the analytical practice. The excellent chemometrics does not raise doubts about the great prospective of the mass spectrometry as a robust analytical instrumentation applicable to 3D structural analysis. In the light of the new contributions to our principle SD concept and model equation (1) consisting of developing of the new model equation (2), it becomes more than obvious that by means of the mass spectrometry, there can be examined not only the multidimensional molecular conformations and electronic structures of analytes, but also can be gained very important insights into the affect of the temperature on the multidirectional molecular structure.

The following reasons bring us to write, as well as, that the two equations (1) and (2) discussed in this work are exact models, describing the temporal behavior of the experimental MS intensity with respect to the scan time of the measurements. This statement is closely connected with a general question under dispute in developing the quantitative MS methodology: Why we use namely stochastic dynamic methods? The answer to the later question and, in general, the point of view, that namely SD methods appear prominent approaches to quantify instrumental MS outcomes are based on the aim at maximizing in the accuracy, the reproducibility, the precision, the selectivity, and the universal applicability of the theoretical methods and models. In what follows, we shall give reason for our choices of the SD methods. In doing so, we will need to define explicitly what is an adequate evidence for the mentioned above statistical parameters or method performances. On our interpretation, the mutual correlation among different experimental parameters and environmental factors is expressed quantitatively by *probability theories* or theories accounting for the *true frequency* or the *true probability*. This means that, when a true value of outcome parameter (in our case, it is the intensity) is hypothesized, there is evaluated the observable results which would converge when there are carried out a series of repeatedly obtained values of an infinite number of measurements [23,36]. As far as, the experimental MS intensity is approximated to concentration of analyte in solution [1–9]; and, in parallel, the concentration limits of detection of the MS methods is at about $\sim 10^{-9}$–10^{-12} mol.L^{-1}, the range of detectable number of particles is at about $\sim 10^{16}$–10^{13}. The fluctuations of this number of particles are obtained on the base on the so–called $(N)^{1/2}$–law [36]. The fluctuation of the intensity, therefore, should be within $(N)^{1/2} \sim 10^8$–10^6. The accuracy of the MS measurements, however, does not account for fluctuations $\sim 10^6$ particles. The SD theories, therefore, describe exactly the experimentally observable MS phenomena. In other words, due to limits of precision of the experiment the predictability of the instrumental outcome or the predictability of the measurable variable is restricted. Taken together, these facts determine that equation (1) is an exact equation. Because of, the Ornstein–Uhlenbeck's approximation to SD processes and its interpretation by Gillespie are, in fact, exact model equations. The same is true for equation (2). In this context, we do try to justify why to evaluate the *reproducibility* and *sensitivity* of the MS outcome *"intensity"* there is needed to use namely our equations (1) and (2). Because of, as aforementioned, our SD concept is based on exact mathematical models, in spite of our approximation (equation (6).)

A final point. The results achieved in references [1–9], together with those reported to reference [23] and the current crucial contribution evidencing empirically the universal validity of the innovative equation (2) dramatically increase in capability of the mass spectrometry as a powerful instrumental method, in fact, for multidimensional structural analysis. Thus, its applicability becomes far beyond the current routine implementation into the analytical practice as a method for quantitative analysis. In addition to, the latter research effort has shown that the application of SD concepts to quantitative methodology of the mass spectrometric methods opens new door for derivation of exact functional relations, describing connections among instrumental outcomes and environmental parameters. Thus, we detail very important multidimensional structural information with respect to a broad spectrum of environmental experimental factors. Owing to the great applicability of the mass spectrometry to different interdisciplinary research areas, explicitly, highlighting the areas not only of the *analytical chemistry* or the *environmental chemistry,* or both, but also other research fields, for instance, the *medicinal chemistry, clinical diagnostics, forensic chemistry, food chemistry, nuclear forensics, pharmacy, archeology, agricultural science,* etcetera, it is more than clear that our methodology based on SD and the derivation of the two model equations (1) and (2) contribute crucially not only to the fields of the Chemistry, but also to many other interdisciplinary research areas.

5. CONCLUSION

Throughout this commentary, in fact, we have provided empirical proof of validity and universal applicability of our new model equation

$$\ln\left(\overline{(I-\bar{I})^2}\right) = -\ln\left(-\left(\ln\left(\frac{k_B \times T}{m}\right)\right)^3 \times \frac{2 \times \Delta t \times T \times k_B}{m \times D_{SD}}\right)$$

, connecting among stochastic dynamic diffusion parameter *"D_{SD}"* introduced *via* our relation derived more recently

$$D_{SD} = 1.3193 \times 10^{-14} \times A \times \frac{\left(\overline{I^2} - \left(\bar{I}\right)^2\right)}{\left(I - \bar{I}\right)^2}$$

( ), the instrumental outcome intensity and the experimental factor *temperature,* respectively. The analysis encompasses eight analytes studied mass spectrometrically under four different ionization methods. In total, fifty seven fragment MS ions are examined. Excellent coefficients of linear correlation between experiment and theory (r = 0.9865$_8$–0.9994$_{77}$) have been obtained. This fact evidences the validity of the two equations.

REFERENCES

[1] B. Ivanova, M. Spiteller, Quantification by matrix–assisted laser desorption ionization mass spectrometry using an approach based on stochastic dynamics. Experimental and theoretical correspondences, GRIN Verlag, München, (2018), pp. 1–86, ISBN: 9783668703179.

[2] B. Ivanova, M. Spiteller, An experimental and theoretical mass spectrometric quantification of non–covalent interactions in high order homogeneous self–associates of

nucleobases and nucleosides, NOVA Science Publisher, New York, 2018, pp. 1–134, ISBN: 9781536142433.

[3] B. Ivanova, M. Spiteller, Experimental and theoretical mass spectrometric quantification of diffusion parameters and 3D structural determination of ions of L-tryptophyl-L-tryptophan in electrospray ionization conditions in positive operation mode, J. Mol. Struct. 1173 (2018) 848–864.

[4] B. Ivanova, M. Spiteller, Experimental mass spectrometric and theoretical treatment of the effect of protonation on the 3D molecular and electronic structures of low molecular weight organics and metal–organics of silver(I) ion, In book: Protonation: properties, applications and effects, A. Germogen (Ed.) (2019), Nova Science Publishers, N.Y., pp. 1–182, ISBN: 978-1-53614-886-2.

[5] B. Ivanova, M. Spiteller, Mass spectrometric experimental and theoretical quantification of reaction kinetics, thermodynamics and diffusion of piperazine heterocyclics in solution, In book: Advances in Chemistry Research, J. Taylor (Ed.), Publisher: NOVA Science Publishers Inc., N.Y., Vol. 48, (2019) pp.1-82, ISBN: 978-1-53614-724-7.

[6] B.Ivanova, M.Spiteller, 3D structural analysis of copper(II) complex of glycine – Experimental mass spectrometric and theoretical quantum chemical approach, J. Mol. Struct. 1179 (2019) 192–204.

[7] B. Ivanova, M. Spiteller, Stochastic dynamic electrospray ionization mass spectrometric diffusion parameters and 3D structural analysis of coordination species of copper(II) ion with glycylhomopentapeptide and its dimeric associates, J. Mol. Liq. 282 (2019) 70–87.

[8] B. Ivanova, M. Spiteller, Electrospray ionization and collision induced dissociation mass spectrometric quantitative conjunctions with the experimental intensity of the analyte ions of metal–organics – stochastic dynamics, GRIN Verlag: München (2019), pp. 1–218.

[9] R. Cole (Ed.), Electrospray ionization mass spectrometry, J. Wiley and Sons (1997) N.Y. pp. 1–577.

[10] R. Cole, A. Harrata, Charge–state distribution and electric–discharge suppression in negative–ion electrospray mass spectrometry using chlorinated solvents, Rapid. Commun. Mass. Spectrom. 6 (1992) 536–539.

[11] R. Cole, A. Harrata, Solvent effect on analyte charge state, signal intensity, and stability in negative ion electrospray mass spectrometry; implications for the mechanism of negative ion formation, J. Am. Soc. Mass Spectrom. 4 (1999) 546–556.

[12] G. Weng, Z. Liu, J. Chen, F. Wang, Y. Pan, Y. Zhang, Enhancing the mass spectrometry sensitivity for oligonucleotide detection by organic vapor assisted electrospray, Anal. Chem. 89 (2017) 10256–10263.

[13] R. Bain, C. Pulliam, R. Cooks, Accelerated hantzsch electrospray synthesis with temporal control of reaction intermediates, Chem. Sci. 6 (2015) 397–401.

[14] T. Nohmi, J. Fenn, Electrospray mass spectrometry of poly(ethylene glycols) with molecular weights up to five million, J. Am. Chem. Soc. 114 (1992) 3241–3246.

[15] S. McLuckey, G. Van Berkel, G. Glish, Reactions of dimethylamine with multiply charged ions of cytochrome C, J. Am. Chem. Soc. 72 (1990) 5668–5670.

[16] J. Fenn, M. Mann, C. Meng, S. Wong, C. Whitehouse, Electrospray ionization-principles and practice, Mass Spectrom. Rev. 9 (1990) 37–70.

[17] C. Meng, J. Fenn, Formation of charged clusters during electrospray ionization of organic solute species, Org. Mass. Spectrom. 26 (1991) 542–549.

[18] J. Iribarne, B. Thomson, On the evaporation of small ions from charged droplets, J. Chem. Phys. 64 (1976) 2287 (9 pages).

[19] B. Thomson, J. Iribarne, Field induced ion evaporation from liquid surfaces at atmospheric pressure, J. Chem. Phys. 71 (1979) 4451 (14 pages).

[20] B. Thomson, J. Irlbarne, P. Dzledzic, Liquid ion evaporation/mass spectrometry/mass spectrometry for the detection of polar and labile molecules, Anal. Chem. 54 (1982) 2219–2224.

[21] (aa) X. Yan, Y. Lai, R. Zare, Preparative microdroplet synthesis of carboxylic acids from aerobic oxidation of aldehydes, Chem. Sci. 9 (2018) 5207–5211.

[22] A. Gallo, Jr., A. Farinha, M. Dinis, A. Emwas, A. Santana, R. Nielsen, W. Goddard III, H. Mishra, The chemical reactions in electrosprays of water do not always correspond to those at the pristine air-water interface, Chem. Sci. (2019) DOI: 10.1039/c8sc05538f.

[23] B. Ivanova, M. Spiteller, Temperature dependence on stochatic dynamic diffusion coefficients of heated electrospray ionization and atmospheric pressure chemical ionization mass spectrometric intensities of analyte ions of configurationally locked polyenes – experiment and theory, Int. J. Mass Spectrom. (2019) submitted.

[24] D. Gillespie, Exact numerical simulation of the Ornstein-Uhlenbeck process and its integral, Phys. Rev. E (1996) 2084–2091.

[25] W. Coffey, Y. Kalmykov, J. Waldron, The Langevin Equation, World Scientific Publishing, N.J. (2004) pp. 1–679.

[26] D. Gillespie, Markov Prosesses, Academic Press (1992) N.Y., pp. 1–566.

[27] G. Uhlenbeck, L. Ornstein, On the theory of the Brownian motion, Phys. Rev. 36 (1930) 823–841.

[28] D. Gillespie, The mathematics of Brownian motion and Johnson noise, Am. J. Phys. 64 (1996) 225–240.

[29] E. Nelson, Dynamic theories of Brownian motion, Princeton University Press, 1967, N.J., pp. 1–142

[30] A. Einstein, Über die von der molekularkinetischen Theorie der Wärme geforderte Bewegung von in ruhenden Flüssigkeiten suspendierten Teilchen, Ann. Phys. 17 (1905) 549–560.

[31] A. Einstein, zur Theorie der Brownischen Bewegung, Ann. Phys. 19 (1906) 371–381.

[32] U. Schwarz, Stochatic Dynamics, Lecture notes (2019) Heidelberg University, pp. 1–78.

[33] D. Gillespie, Fluctuation and dissipation in Brownian motion, Am. J. Phys. 61 (1993) 1077– 1083.

[34] T. Veenstra, Electrospray ionization mass spectrometry in the study of biomolecular non-covalent interactions, Biophys. Chem. 79 (1999) 63–79.

[35] K. Song, R. Spezia, Theoretical mass spectrometry, De Gruyter (2018) Berlin Boston, pp 1–228.

[36] P. Schuster, Stochasticity in Processes, Fundamentals and Applications to Chemistry and Biology, Springer Verlag (2016) Berlin Heidelberg, pp. 1–718.